Ederson Dias da Rocha
Daniel Mantovani
Aline A. Baptista

Public space and air quality: sick building syndrome

AF671811

Ederson Dias da Rocha
Daniel Mantovani
Aline A. Baptista

Public space and air quality: sick building syndrome

Sick Building Syndrome (SBS)

ScienciaScripts

Imprint

Any brand names and product names mentioned in this book are subject to trademark, brand or patent protection and are trademarks or registered trademarks of their respective holders. The use of brand names, product names, common names, trade names, product descriptions etc. even without a particular marking in this work is in no way to be construed to mean that such names may be regarded as unrestricted in respect of trademark and brand protection legislation and could thus be used by anyone.

Cover image: www.ingimage.com

This book is a translation from the original published under ISBN 978-613-9-68194-5.

Publisher:
Sciencia Scripts
is a trademark of
Dodo Books Indian Ocean Ltd. and OmniScriptum S.R.L publishing group

120 High Road, East Finchley, London, N2 9ED, United Kingdom
Str. Armeneasca 28/1, office 1, Chisinau MD-2012, Republic of Moldova, Europe
Printed at: see last page
ISBN: 978-620-8-19670-7

Copyright © Ederson Dias da Rocha, Daniel Mantovani, Aline A. Baptista
Copyright © 2024 Dodo Books Indian Ocean Ltd. and OmniScriptum S.R.L publishing group

SUMMARY

SUMMARY

In recent decades, people have spent most of the day indoors with little or no air exchange with the outside environment, increasing the use of air conditioning techniques in workplaces and homes, creating conditions that are potentially harmful to human health. Libraries are places where people circulate a lot and where there is a huge collection of books, but there have been few studies on this subject. The library of the State University of Maringà was therefore selected as the setting for this study. The aim of this study was to observe the quality of indoor air and to relate closed environments with natural ventilation and artificial ventilation by air conditioning equipment to the Sick Building Syndrome (SED) by relating the possible symptoms presented by workers in the area such as allergies, irritation and respiratory problems. The building studied has an area of 13,298.03 m^2 divided into a basement, first floor and two floors. The lower floor makes use of both natural ventilation through windows and a split-system air conditioning system, while the second floor has only natural ventilation through windows. The study used observations, data recording and questionnaires. The questionnaire showed that the workers manifested a large number of typical symptoms related to EDS. This highlights the need for corrective action to be taken on air conditioning equipment in order to maintain the quality of the air in the environment in comfortable conditions so as to guarantee the health and well-being of the occupants. It can be concluded that a library has favorable conditions for the proliferation of biological agents, especially fungi, which can occur due to the large circulation of people, the collection of books, humidity, temperature and aggravated when the preservation and maintenance measures of the air conditioning systems are insufficient to preserve the quality of the internal air, which can lead to health problems, a drop in productivity and the well-being of the people who use the place.

Key words: Indoor air quality, air conditioning, biological agents, sick building syndrome, library.

1 INTRODUCTION

Studies by the Intergovernmental Panel on Climate Change (IPCC) point to an increase in the Earth's average temperature, caused by an increase in greenhouse gas emissions. Environmental discussions are no longer just problems for future generations and are now evident in people's daily lives. Economic and cultural transformations have significantly altered society's way of life, with migration from rural areas to cities, increasing the emission of pollutants.

With the verticalization of large urban centers, people tend to spend most of the day in offices, factories or even in their homes, in closed environments without a continuous exchange of natural ventilation, which has intensified the adoption of air conditioning equipment for thermal comfort.

Man's own activities can alter the physical, biological or chemical characteristics of the air inside buildings, causing contamination of environments by building materials, cleaning products, mold and domestic activities. These actions generate various contaminants such as carbon monoxide and dioxide, ammonia, sulphur oxide and nitrogen. Contributors to indoor air contamination include aerodispersoids (dust and fibers); biological agents such as bacteria, viruses, fungi, pollen grains, dust mites and spores; chemical contaminants such as volatile organic compounds (VOCs) and formaldehyde; substances generated by human metabolism; inadequate ventilation, among others.

As for unsatisfactory air quality conditions, children and the elderly are the most sensitive group to loss of air quality. Micro-organisms present in these places can cause allergic reactions, the symptoms of which include sneezing, watery eyes, coughing, respiratory impairment and fever, as well as being etiological agents of pneumonia, rhinitis and asthma. The main diseases associated with biological pollutants are Legionnaires' disease caused by the bacterium *Legionella spp*, humidifier fever, bronchial asthma, allergic pneumonitis or extrinsic alveolitis and pneumonia.

Inadequate maintenance of indoor air can lead to health problems and the proliferation of harmful micro-organisms, as well as reducing the efficiency of air conditioning equipment and increasing energy costs. Air conditioning is no longer seen as a futility, but as an element that adds value to the enterprise by providing more comfort in the workplace and increasing the efficiency of workers.

Due to the risks related to the health of the population, there was a need to create regulations for the quality of air in artificially air-conditioned environments. In 1976, the Ministry of the Interior issued Decree No. 231, which established technical criteria for assessing total suspended particles in the air. Other federal bodies such as the National Environment Council (CONAMA), the National Health Surveillance Agency (ANVISA) and the Brazilian Technical Standards (NBR) also provided support and guidelines for maintaining indoor air quality.

1.1 Objective

1.1.1 General Objective

To discuss the aspects that contribute to air contamination in closed environments, such as the library of the State University of Maringà, in terms of physical, chemical and biological agents and to analyze the possible impacts on human health and the characterization of Sick Building Syndrome (SED).

1.1.2 Specific Objectives

The specific aim of this study was to identify the complaints and health symptoms that workers at the library of the State University of Maringà, Maringà Campus, have according to the Sick Building Syndrome (SED).

- Survey procedures, laws and technical standards that highlight the importance of preserving and monitoring air quality in closed and artificially ventilated environments;
- Checking artificial ventilation systems in closed environments, monitoring contaminants, as well as identifying factors that interfere with the comfort and health of occupants in environments such as libraries;
- To apply a questionnaire that can contribute to the studies, identifying possible complaints and health problems of workers related to Sick Building Syndrome (SED);
- Analyze the results obtained through the "in loco" questionnaire of the selected environment and compare the parameters of the SED;
- To discuss the importance of monitoring air quality in artificially air-conditioned environments and technical procedures for the prevention and maintenance of air-conditioning equipment.

1.2 Justification

Given the importance of monitoring closed environments, there was an interest in studying the legislation and technical standards created to control the air quality of these environments and preserve the health of their occupants who spend most of the day in offices, schools, hospitals and other environments where air conditioners are used on a large scale.

Factors such as contaminants, diseases, preventive and corrective actions will be analyzed in order to ensure greater efficiency in the maintenance processes of air conditioners, with the aim of recognizing the agents of contamination of closed environments in the library of the State University of Maringâ (UEM) on the Maringâ campus.

1.3 WORK METHODOLOGY

The research has a mixed approach and is a case study of air quality in closed and artificially air-conditioned environments on the premises of the UEM Maringâ campus library. It was qualitative in nature, producing data from observations through bibliographical reviews and visits to the areas in order to gain a broader understanding of the effects on human health of exposure to contaminating agents. An anonymous questionnaire was also administered to the workers with the aim of obtaining a diagnosis of the comfort of the environment and satisfaction with the library's Indoor Air Quality (IAQ) in terms of individual perception.

2 THEORETICAL BACKGROUND

2.1 The Emergence of Cities and the Impairment of Air Quality

The city was born as a village, formed when people who didn't cultivate the land began to work in industries, becoming the driving force of evolution and changing rapidly. More than four thousand years ago, Sumerian cities were already surrounded by walls and moats and were no longer open in natural environments, but had artificial landscapes. The world changed with the industrial revolution, with a reduction in infant mortality generating population growth and an increase in the consumption of goods and services, achieving economic and technological development, so cities quickly developed in contrast to rural environments (BENEVOLO, 2011).

For the same author, from 1948 onwards, a new city model was adopted, idealized by the bourgeoisie, establishing regulations, a period that marked the transition from liberal to post-liberal cities. This change was soon successful and reshaped the great cities of Europe and allowed new ones to be founded. This reorganization influenced the structures of the cities we live in today.

With regard to migration in Brazil, the population that left the countryside for urban centers grew by 29.5% between 1965 and 1994, but from 2000 to 2010 the average fell to 0.65%, reaching 84.4% of the Brazilian population (IBGE, 2010). This movement has brought problems to society, causing major congestion on public roads with vehicles, contributing to the deterioration of the air in cities, also damaging indoor air, so artificial ventilation for thermal comfort and ensuring better air quality, began to become more popular (BRICKUS; AQUINO, 1999).

2.2 History of Refrigeration

The history of refrigeration begins even before Christ, with its first reports in China, where ice from rivers was preserved in straw-covered pits to be used in the hottest periods to improve the taste of drinks. In the 18th century, refrigeration equipment and artificial ice-making equipment were developed, but it was only in the 20th century, with the discovery of the cryogenic properties of gases, that refrigeration for preserving food became widespread (DI RIENZO, 2006).

Willis Haviland Carrier was the pioneer of studies to control indoor environmental conditions. The engineer began his first scientific studies into air conditioning in 1902, in New York City, at the service of *the Sacket Wihems Lithography and Publishing* company, which was trying to solve a problem in the printing of papers that suffered interference in their quality due to high humidity and temperature. Thus, Carrier proposed what would be recognized as the world's first mechanical air conditioning process, designing a machine that made it possible to circulate air through artificially cooled ducts, reducing the humidity and temperature of the environment, meeting the needs of the production process (CARRIER DO BRASIL, 2013).

At random, Carrier made another discovery, observing the importance of the dew point while waiting for a train at Pittsburg station, formulating the theory of air conditioning based on temperature control, humidity, circulating air and air purification (DI RIENZO, 2006).

According to Willis Carrier's biography (2013), in 1903 Carrier patented what would become the first air treatment device, a *spray-type* device that could wash, humidify or dehumidify the air. In 1913 he created the first air humidification device, capable of serving rooms such as offices and laboratories. In 1914, *Pittsburgh's Allegheny General* Hospital reduced the mortality of babies from dehydration and respiratory problems by adopting a system in nurseries that increased the temperature and humidity of the air.

Carrier's discoveries spread around the world and his studies are still references today (ARAÙJO, 2011). Carrier's air conditioning equipment revolutionized dairies, refrigerators, homes, shopping malls, large buildings and libraries. In 1998, *Times Magazine* recognized Carrier as one of the 100 most influential people of the 20th century (CARRIER DO BRASIL, 2013).

2.3 Man and Thermal Comfort

Man is a homeothermic animal, with a body temperature of 37°C, with a lower limit of 32°C and a maximum upper limit of 42°C for survival. The homeothermic organism can be compared to a thermal machine where its energy is obtained through thermal phenomena, also called metabolism, which is produced by the human organism and occurs through internal chemical reactions, such as the combination of carbon and oxygen (FROTA; SCHIFFER, 2016).

Thermal comfort is understood as the environmental conditions that guarantee, among other factors, thermal and anthropometric well-being, as well as air quality. Thermal comfort is reflected in people's satisfaction with the environment they occupy, in a balance between the thermal exchanges that the human body is subjected to, when the result of these exchanges is canceled out, thermal comfort orbits (LAMBERTS; DUTRA; PEREIRA, 2013).

The skin is the main organ responsible for heat exchange, which is regulated by blood flow and rises when this flow increases, so when a person feels thermal discomfort their mechanism is vasomotor regulation through vasodilation and active perspiration through sweat glands (FROTA; SCHIFFER, 2016).

The transfer of heat from the human body to the environment occurs through sensible heat released by conduction, irradiation and convection, the latter of which is responsible for the temperature of the air surrounding the body. When heated, this air moves upwards and gives way to a new layer (SILVA, 2002).

According to Frota and Schiffer (2016), when heat is lost to the environment through dry exchange, the process is called sensible heat, while heat lost through wet exchange is known as latent heat. The conditions for heat exchange to occur are between bodies at different temperatures and a change in the state of aggregation.

Ruas (2001), Lamberts, Dutra and Pereira, (2013) emphasize that environmental conditions have as factors the average radiant temperature, relative humidity and air temperature, as well as the clothing used and the types of physical activities performed.

Thermal comfort can also be affected by other factors as shown in Figure 1, involving air circulation, radiation from neighboring surfaces, odors, dust and noise, as well as personal factors such as age, weight, activities performed and the type of clothing worn (PIRANI, 2001).

Figure 1: Factors affecting thermal comfort. Source: PIRANI, 2001.

Silva (2002) reported that as the air temperature in the environment rises, the rate of heat exchange decreases. When the normal body temperature reaches around 37°C, the rate of exchange is canceled out. If the air temperature rises above 37°C, the body absorbs heat from the warm air and when the reverse occurs, the flow of heat transfer is from the body to the environment. The human body's process between latent and sensible heat, when it occurs under normal conditions by varying temperature and humidity, is known as homeothermic equilibrium.

The interactions between temperature, humidity and air are of great importance in maintaining IAQ. They are preponderant factors for the proliferation of chemical and biological agents, directly affecting people's comfort and health conditions (MARTINEZ, CALLEJO, 2006).

The feeling of comfort is subjective and can vary for each person, which is why an environment can be considered pleasant for one person and unpleasant for another, so it can be recognized that the best environmental conditions of comfort are those that meet the needs of the greatest number of people in a given place (RUAS, 2001).

2.4 Air conditioning, ventilation and refrigeration in closed environments.

In Brazil, the first air conditioning system was installed in the Morro Velho gold mines at the end of the 1920s, by the *General Electric* company. The system was adopted to improve the conditions of workers extracting ore at great depths. The first installation by a genuinely Brazilian company took place at the Jockey Club in Rio de Janeiro, but was questioned by the population, as it was believed that it could pose a health risk, causing flu and colds (DI RIENZO, 2006).

According to Pena (2002), the process of thermally adapting an internal environment can vary according to specific objectives, and can be divided into air conditioning, refrigeration and ventilation systems:

- Air conditioning: This is based on air treatment, controlling temperature, internal pressure, air purity and humidity, as well as adjusting the temperature to around 20°C. It can be divided into air conditioning for environmental comfort in homes, offices and businesses or industrial air conditioning to improve processes such as in the weaving, printing and electronics industries.

- Refrigeration: This method is widely used in cold rooms or refrigerated counters, where air is used as a fluid and cooled to temperatures around 0° C to -10°C.

- Ventilation: Air is used to control the temperature of the room, removing the internal thermal load and renewing it with the outside air. It is also used to remove pollutants and odors, but the temperature cannot be controlled to a fixed value.

According to Silva (2002), all places have ventilation, either directly or indirectly, and it is present in all the environments that people frequent and work in. Refrigeration can be classified into three branches: Commercial, Industrial and Scientific.

According to the same author, the most common type of refrigeration is commercial, which is used in homes, offices and cinemas, where the temperature is kept between 20 and 26°C for people's thermal comfort. Scientific refrigeration is used in the areas of science and research, and is widely used in laboratories.

A refrigeration system must control four parameters that can be assessed directly or indirectly: air temperature, temperature of the surrounding surfaces, air humidity and air velocity (PIRANI, 2001).

When an environment has a high temperature, humidity or saturated atmosphere, the air is less receptive to sweat. The refrigeration solution is to exchange the air in the environment for cooler, drier air, which will increase the air's capacity to absorb sweat, where the air will act like a sponge, absorbing the humidity until it is saturated, after which it will no longer be able to retain any more liquid. So there is equipment that performs this control mechanically (SILVA, 2002).

2.5 Air conditioners

Air conditioners are systems used to treat air in order to control the temperature, relative humidity, purity and distribution of air in closed environments, generating thermal comfort for people and special conditions for equipment and processes. As well as being used to make the environment more pleasant by heating or cooling enclosed spaces, there are environments such as laboratories, hospitals and data processing centers, among others, where the use of air conditioning is essential (ARAÙJO, 2011).

According to the same author, the most commonly used units for calculating the power of air conditioning equipment are the British *Thermal* Unit (BTU) and the Refrigeration Ton (TR), where 12,000 BTU corresponds to 1 TR. When sizing air conditioning equipment with adequate power to cool a particular room, factors such as the area to be cooled, the number of people occupying the room, the number of windows and doors, whether there is electrical equipment and the exposure of the room to the sun must be taken into account.

Air conditioners can be classified into comfort and process systems. The comfort system is used to treat the air to ensure the health of the occupants, while the process system is used to control product manufacturing conditions (CORRÊA, 2010).

Air conditioners use the evaporation of refrigerants to filter, cool and dehumidify the air, as well as renewing the air in the room and heating it. Air conditioning systems can be classified into direct expansion used in smaller environments and indirect expansion used on a large scale in industries, businesses and hospitals (ALBERICO, 2003).

2.6 Air Conditioning Systems Maintenance

In order to guarantee air quality in indoor environments that use air conditioning systems, regular maintenance must be carried out on filters, turbines and equipment coils, which end up retaining fungi, bacteria, dust mites and other agents that can cause illness in the occupants of these areas. When maintenance is carried out carefully and regularly, it helps to improve the quality and useful life of the equipment, reduces energy waste, improves people's health, increases productivity and influences the institution's costs (ARAUJO, 2011).

There are different ways of classifying maintenance methods, the most common of which are corrective, preventive and predictive (PINTO; XAVIER, 2009):

- Corrective: Carried out when a fault *has already* occurred, the cost of maintenance can be cheaper than preventive maintenance, but it results in unscheduled downtime and reduced process efficiency. To avoid greater damage and increase the response time to the fault, it is important to maintain trained professionals and spare parts for emergency interventions.

- Preventive: This is the most recommended type of maintenance to be adopted, with periodic inspections aimed at carrying out interventions such as renovations and replacement of parts. Compared to corrective maintenance, it is more expensive as parts that still work are replaced, but it reduces unexpected failures and allows for scheduled downtime, which in the end can be more economical than corrective maintenance.

- Predictive: This is considered one of the elements of preventive maintenance, as it also has the purpose of inspecting equipment. This type of maintenance aims to predict the life limit of parts and components and thus extend the interval between interventions, allowing replacement to be optimized. Experts say that predictive maintenance is considered to be the most efficient compared to other methods, as it uses more advanced technologies and must be carried out by an even more specialized workforce.

Preventive or scheduled maintenance is based on the equipment manual, where the services to be carried out and the frequency at which the equipment should be monitored are already planned in order to reduce consumption and reduce problems such as unwanted downtime (PENA, 2002).

Table 1 refers to the frequency of cleaning and inspections of air conditioning systems established by Resolution No. 9 (ANVISA, 2003).

Table 1: Frequency of cleaning and maintenance procedures for system components:

Component	**Frequency**
External air intake	Monthly cleaning or when disposable until obliteration (maximum 3 months)
Filter units	Monthly cleaning or when disposable until obliteration (maximum 3 months)
Condensate tray	Monthly
Heating coil	Half-yearly descaling and quarterly cleaning
Cooling coil	Half-yearly descaling and quarterly cleaning
Humidifier	Half-yearly

Mixing plenum/machine room	Half-yearly
Inspection	Half-yearly

Source: RESOLUTION No. 9 (ANVISA, 2003).

One of the factors that encourages the proliferation of fungi, viruses, dust mites and bacteria in indoor environments is related to the incorrect cleaning of filters and cooling ducts, which can cause respiratory, infectious or allergic diseases to the occupants of the room (CARTAXO. *et al.* 2007).

The cleaning of more restricted areas such as ducts can be carried out by robots, equipped with a camera that films the internal area, guiding the operator precisely to the most critical areas to clean, identifying or not sources of contamination (ARAÙJO, 2011).

There are special filters used as the main method of disinfecting air for micro-organisms, they are low cost and easy to handle, but they do not completely destroy these micro-organisms, as they only transfer them from one environment to another, making the process more expensive and requiring proper disposal (MONTAGNER; PASCHOALINO; JARDIM, 2005).

Resolution no. 9 (ANVISA, 2003) recommends that environments with air conditioning systems with a capacity of 5 TR or more (15,000 Kcal/h = 60,BTU/h) should have a technical manager in accordance with Ordinance GM/MS no. 3.523/98, who will be responsible for carrying out biological, chemical and physical assessments of the environments. They will also have to make adjustments to the air conditioning system when the requirements of Resolution No. 9 are not met, and provide a record of the information, results of the assessments and maintenance, disclosing them to the occupants of the environments.

Law No. 13,589 (BRASIL, 2018) was recently passed, requiring the preparation of a Maintenance, Operation and Control Plan (PMOC) for all public buildings and buildings for collective use that have artificial air conditioning systems.

2.7 Indoor Air Quality (IAQ)

Indoor air is understood to be that present in homes, offices, schools and hospitals, not including industries. Addressing air quality in these environments is important because several factors can influence it. Ventilation stands out as the main one, but man himself in his activities also contributes to the deterioration of indoor air (SCHIRMER, W.N. *et al.* 2008).

In tropical regions, the use of air conditioning is essential for thermal comfort. The growth of cities and industrialization have contributed to atmospheric pollution, although there are few studies that relate urbanization and the climate of cities, it is known that densely populated areas, vehicle traffic and deforestation also harm IAQ and consequently human health (CARTAXO. *et al.* 2007).

There are various interactions between internal and external environments that can influence people's health, between natural elements and those created by man himself. These different elements can be separated into physical, chemical and biological agents, as well as psychological, social and cultural ones (MARTINEZ, CALLEJO, 2006).

The study of Indoor Air Quality (IAQ) is considered to be new. It emerged in the 1970s, when the scarcity and rationalization of energy created environments with low internal and external air exchange, and in order to obtain greater efficiency in cooling and heating systems, buildings were constructed to obtain greater thermal sealing. These environments contain various types of chemical substances that can be dispersed into the air from lining products, finishes, furniture and the outside air, increasing complaints about air quality, especially in cold climate countries (GIODA; NETO, 2003).

The scientific interest in IAQ is due to the advances that have made it possible to better understand the mechanisms that can cause negative effects on human health, enabling new strategies to diagnose, control and prevent diseases caused by IAQ (GRAUDENZ; DANTAS, 2007).

According to Brickus and Neto (1999) and Schirmer *et al.* (2008) people spend most of their lives in sealed buildings, motivated by aesthetic issues and thermal comfort, there has been an increase in cases of health problems, arousing the need for studies that point to the increase in the concentration of chemical and biological pollutants in enclosures, justifying the concern about the concentration of air contaminants.

It is estimated that 20 to 30% of people working in office buildings suffer from IAQ symptoms, generating an estimated annual cost of 22 billion dollars (BOECHAT; RIOS, 2011).

In developed countries, there has been a large increase in complaints about air quality, and studies have indicated that in some cases the air quality in enclosed spaces such as homes and offices can be more polluted than the air outside in industrial areas, which is called Sick Building Syndrome (SED). Studies carried out by Brickus and Neto (1999), Martinez and Callejo (2006) mention smoking, burning fuels and using air conditioning, among others, as the main sources of IAQ contamination, as can be seen in Table 2.

Table 2: Typical sources of indoor air pollution:

Environment	Sources	Pollutants
Residence	Smoke	Respirable particles, CO, VOCs, nicotine, PA, phenols, nitrosamines, NO2,

	Gas stove	NO2, CO, gaseous hydrocarbons
	Wood-burning stove	Respirable particulates, CO, HPA, NO2.
Residence	Wet surfaces and materials	Biological agents.
	Cleaning activity	Respirable particles, VOCs.
Office	Building materials and furniture	Formaldehyde, VOCs, radon.
	Air conditioning	Biological agents, outside air.
	Photocopiers and laser printers	VOCs, respirable particles and ozone.
	Cleaning activities	Respirable particles, VOCs.
Transportation	Smoke	Respirable particles, CO, VOCs, nicotine, PA, phenols, nitrosamines, NO2,
	Burning fuel	Particulate matter, CO, HPA, NO2, VOCs, aldehydes.

Source: BRICKUS and NETO, 1999.

2.7.1 Sick Building Syndrome (SED)

In 1982, the World Health Organization (WHO) defined the Sick Building Syndrome (SED) as a public health problem, which consists of the state in which people who occupy an environment have a high level of symptoms related to poor air quality, these symptoms can be headaches, problems with the eyes, nasal passages, throat, chest, fatigue, lethargy, skin abnormalities and lack of concentration at work (COSTA; COSTA, 2006).

The WHO classifies buildings as sick when they have the following characteristics (MARTiNEZ; CALLEJO, 2006):

- 20% of indoor occupants report complaints and health problems caused by poor air quality.
- Rooms equipped with forced-air ventilation and air conditioning systems, even if they also have natural ventilation.
- When the surfaces of furniture, walls and floors are found to be covered in textiles or other elements originating from the occupants.
- Closed and relatively warm environments, with little or no access to the outside environment.
- It often occurs in fast-build, low-cost buildings.

As mentioned by Martinez and Callejo (2006), the WHO divided sick buildings into two types in 1982:

- Temporarily unhealthy buildings: these are buildings that have just been built or are undergoing renovation, where the people occupying the environment show symptoms related to EDS, but these disappear over time, on average after six months.

- Permanently sick buildings: these are buildings in which the symptoms of EDS are permanent for the occupants, even if corrective measures are taken in the buildings, the symptoms do not disappear over time.

Outdoor air quality is a major factor in IAQ, with cars and construction activities being the most common sources of contamination. Agents have access to indoor environments through doors, windows and air conditioning equipment (MARTiNEZ; CALLEJO, 2006).

EDS is known around the world for impairing the performance of workers who remain in areas with low air exchange. Among the main symptoms is malaise, which reduces productivity (MONTAGNER; PASCHOALINO; JARDIM, 2005).

Kavgic *et al.* (2008) reports that various symptoms have been related to people who work indoors, such as headaches, irritation of the eyes, nose and throat, sneezing, runny nose and so on. Reaffirming a study already discussed by Sterling, Collett and Rumel (1991) who highlight a study carried out by Hedge in 1986, which looked at a population of 796 people who worked in three closed buildings and another 214 workers who stayed in two buildings with natural ventilation. The occupants of the buildings answered questionnaires and were asked if they had any symptoms related to EDS during the working day. They could choose between four categories: never, rarely, sometimes and always. The aim of the survey was to compare EDS symptoms between occupants in an environment with natural and artificial ventilation, as shown in Table 1.

Table 1: Comparison of health complaints between two types of office buildings:

Symptom	**Ventilation Air conditioning (%)**	**Natural ventilation (%)**
Drowsiness	69,2	44,5
Fatigue	68,0	52,4
Pain and headache	67,2	50,5
Ocular impression	52,1	45,9
Cold	50,2	32,4

Sore throat	47,9	28,3
Nasal impression	45,5	26,5
Difficulty focusing	42,9	28,8
Back pain	41,8	41,4
Neck pain	41,2	39,5
Cold extremities	40,7	38,8
Dry skin	29,9	16,7
Depression	25,1	25,2
Dizziness	23,6	15,5
Muscle pain	21,1	17,2
Weakness	20,3	9,1
Nausea	19,4	7,8
Difficulty breathing	12,2	5,7
Chest wheezing	9,8	6,8
Fever	8,1	2,0

Source: HEDGE; STERLING; STERLING (modified), 1986.

Symptoms related to EDS do not affect all occupants of the indoor environment, and when they do manifest themselves they do not disappear when the person leaves the place. Identifying and removing the sources of exposure can help to minimize the impact on occupants' health and reduce complaints (COSTA; COSTA. 2006).

In 1993, the most complex study on IAQ in offices was carried out, looking at complaints in the workplace such as headaches, respiratory and eye problems, among others. The study found that complaints about air quality were twice as high among people who worked in environments with air conditioning than those with natural ventilation (GRAUDENZ; DANTAS, 2007).

Humans are one of the main agents of IAQ contamination. As a result of their activities, indoor air can be compromised, as described in Table 3, which highlights the most frequently found compounds (GIODA; NETO, 2003).

Table 3: Indoor air pollution agents:

Pollutant	Largest source of emissions
Carbon dioxide	Metabolic activity, combustion activities, motor vehicles in a garage.
Carbon monoxide	Burning fossil fuels like gas or kerosene, stoves, cigarette smoke.
Formaldehyde	Building materials and furniture.
Volatile Organic Compounds (VOCs)	Adhesives, solvents, building materials, volatilization, combustion, paint, cigarette smoke, cleaning activities, photocopiers, laser printers.
Particles	Re-suspension, cigarette smoke, combustion products, cleaning activities.

Source: GIODA and NETO, 2003.

Thirty of the five hundred types of volatile organic compounds found in indoor air are considered carcinogenic and mutagenic. Benzene and four thousand other compounds found in cigarette smoke are the main sources of non-occupational contamination (COSTA; COSTA, 2006).

Studies show that people tend to manifest symptoms related to EDS in the afternoon, with public environments being more prone to contamination than private ones and complaints being more frequent in buildings with a large circulation of people (MARTiNEZ; CALLEJO, 2006).

In Brazil there are few studies and records on IAQ and EDS, the subtropical climate means that the use of refrigeration does not need to be constant, which may justify the lack of research on the subject in the country. With the need to use central air-conditioning systems due to new trends in the construction of increasingly sealed buildings, it is estimated that Brazil may have an increase in cases of EDS, exposing people to health problems caused mainly by biological agents (GIODA; NETO, 2002).

2.8 Biological Agents Air Contaminants

According to the authors Carmo and Prado (1999) and Martinez and Callejo (2006), the biological contaminants present inside buildings can come from a variety of sources, including the outside air, water reservoirs and mainly from plants, people and animals that release fungi and bacteria.

In closed environments, humans are often exposed to infectious agents transmitted by other people, and contact with microorganisms occurs through internal and external epithelial surfaces. The main route of contamination for airborne microorganisms is the epithelium lining the respiratory system

(MURPHY, 2014).

The biological agents most studied in IAQ literature are bacteria, fungi and dust mites, which mainly affect people with allergies or low immunity (COSTA; COSTA, 2006).

Table 4 shows the main biological agents, their sources of contamination and corrective measures.

Table 4: Possible sources of biological pollutants:

Biological agents	Main indoor sources	Main corrective measures indoors
Bacteria	Stagnant water tanks, cooling towers, condensate trays, dehumidifiers, humidifiers, air conditioner coils and wet and hot surfaces.	Clean and maintain cooling towers; sanitize condensate reservoirs and trays or maintain continuous treatment to eliminate sources; eliminate infiltration; sanitize surfaces.
Fungi	Damp environments and other sources of fungal multiplication, such as damp organic porous materials, damp ceilings, walls and insulation; outside air, inside unmaintained air conditioners and ducts, earthen pots with plants.	Correct environmental humidity; keep leaks, infiltrations and water condensation under strict control; sanitize rooms and components of the air conditioning system or maintain continuous treatment to eliminate sources; eliminate contaminated porous materials; eliminate or restrict potted plants growing in soil, or replace them with water (hydroponics); use G-1 filters when renewing outside air
Protozoa	Contaminated water tanks, trays and humidifiers for unmaintained air conditioners.	Sanitize the reservoir or maintain continuous treatment to eliminate sources.
Virus	Human host.	Adjusting the number of occupants per m2 of area by increasing air renewal; avoiding

		the presence of infected people in air-conditioned environments.
Algae	Cooling towers and condensate trays.	Sanitize condensate tanks and trays or maintain continuous treatment to eliminate sources.
Pollen	Outside air.	Maintain filtration in accordance with ABNT NBR-6401
Arthropods	House dust.	Sanitize fixed surfaces and furniture, especially those covered with fabrics and carpets; restrict or eliminate the use of these coverings
Animals	Rodents, bats and birds	Restrict access, control rodents, bats, bird nests and their droppings.

Source: RESOLUTION No. 9 (ANVISA, 2003).

Coutinho, Ramos and Bortoletto (2002) discussed a case that occurred in 1996 at the Manguinhos library in the city of Rio de Janeiro, where outbreaks of fungi were found contaminating the literary collection, proliferating rapidly and uncontrollably, as there were very favorable temperature and humidity conditions. In addition to environmental factors, the air conditioning system, which was under maintenance, was also a major factor in the contamination process.

The same authors describe the actions taken at the Manguinhos library to contain the contamination, including the closure of the site, the hiring of a company specialized in diagnosing IAQ contamination and a thorough investigation by the institution's board of directors, which detected the presence of *Aspergillus* and *Penicillium*, low-toxicity fungi that cause respiratory and dermal allergies. After the library was closed, evaluations showed a significant reduction in the fungi, attributed mainly to the absence of human occupation, improvements in the ventilation system and interventions in the areas of infestation, which allowed the library to reopen after four months of closure.

Indoor areas, with low air exchange and which use air conditioning to heat or cool the environment, contribute to the proliferation of pathogenic micro-organisms, which can harm human health. More than 60 pathogens such as bacteria, viruses and fungi have been catalogued, some of which, as shown in Table 5, use the air as a means of transportation and form bioaerosols, which can cause allergic reactions, infections and even be fatal to human beings (MONTAGNER; PASCHOALINO;

JARDIM, 2005).

Table 5: Airborne infectious organisms:

Organization	Type
Corynebacterium diphtheriae	Bacteria
Legionella pneumophila	Bacteria
Mycobacterium tuberculosis	Bacteria
Pseudomonas aeruginosa	Bacteria
Staphylococcus aureus	Bacteria
Staphylococcus epidermidis	Bacteria
Adeno Virus Type III	Virus
Coxsackie A2	Virus
Influenza	Virus

Source: MONTAGNER; PASCHOALINO; JARDIM, 2005.

2.9 Biological Agents and Building-Related Diseases (BRD)

The health damage caused by IAQ can manifest itself immediately or even years after exposure. Repeated exposure to a particular pollutant can make a person more susceptible to health problems (COSTA; COSTA, 2006).

Among the symptoms and illnesses caused by the presence of micro-organisms in the air, those of respiratory origin such as pneumonia, rhinitis, asthma and coughing stand out. Allergic reactions such as eye irritation and sneezing and digestive problems are also common (SCHIRMER, 2011).

EDS-related illnesses can present a variety of symptoms and discomfort, many of which are still unknown and vary according to environmental exposure and the susceptibility of each person. Even though some symptoms are not considered serious, they cause loss of productivity and working time (GRAUDEZ; DANTAS, 2007).

Pigeon droppings can also compromise the IAQ, since when they are sucked up by the refrigeration system and projected into the internal environment, bird droppings are a source of contamination with the fungus *Cryptococcus neofrmans*, the infection occurring mainly in the respiratory and central nervous systems (CARTAXO. *et al.* 2007).

Immunodeficient groups are the most affected by the action of *C. neoformans, and* the increase in cases of cryptococcosis is also related to the increase in these groups, such as people with Acquired

Immunodeficiency Syndrome (AIDS), leukemia, transplant patients, tumor patients and patients who make prolonged use of corticosteroids and antibiotics (REOLON; PEREZ; MEZZARINI, 2004).

There are diseases that are directly related to biological pollutants in indoor environments, the main ones being Legionnaires' disease, humidifier fever, bronchial asthma, allergic pneumonitis and lung infections caused by bacteria, viruses and fungi as specified in Table 6 (SCHIRMER. *et al.* 2011).

Table 6: Specific diseases related to buildings:

Pathology	**Etiological agent**	**Pathophysiology**	**Clinical presentation**	**Diagnosis**
Legionellosis	Gram-negative bacteria Legionella pneumophilia.	Respiratory infection caused by viable bacteria dispersed in the air.	Dry cough, fever, muscle pain, chills. It can develop into a 10% to 15% fatality rate.	Identification of the focus and increased titration of IgG antibodies to the same serotype of bacteria as the focus.
Hypersensitivity pneumonitis	Antigens associated with fungi, bacteria, protozoa, insects and endotoxins.	Inflammatory pulmonary reaction with granulomas and fibrosis caused by hypersensitivity reactions.	Acute: fever, myalgia, cough and recurrent pneumonia. Chronic: pulmonary fibrosis	Clinical history, spirometry, lung diffusion, X-rays, antibody dosages and bronchoprovocations.
Humidifier fever	The same as hypersensitivity pneumonitis	Infamous and uneducated response of glands.	Virus-like illness with fever, myalgia and asthenia, without pulmonary symptoms or sequelae.	Clinical diagnosis associated with environmental findings.
Rhinitis and asthma	Mites, fungi, animal allergens, bacterial toxins, temperature differences.	Nonspecific irritation or exacerbation of pre-existing allergic	Nasal itching and obstruction, runny nose, sneezing, coughing,	High exposure to positive skin or RAST tests for occupational allergens, provocation

		inflammation.	wheezing and dyspnea.	
Eczema or contact dermatitis	Exposure to irritating products such as glass ware, cleaning products, stationery, unknown.	Irritant (mostly) or allergic contact dermatitis.	Itching, scaling, erythema, papules or vesicles.	Environmental analysis, clinical history and contact tests for non-toxic substances.

Source: GRAUDEZ and DANTAS, 2007.

There is a significant difference between ERD and EDS, even though both are characterized by causing health problems to indoor occupants, the symptoms of EDS tend to disappear when the occupant is away from the contaminated environment for a certain period, while the symptoms of ERD will not necessarily disappear (SPENGLER; SAMMET; MCCARTHY, 2001).

The human body itself creates defenses to prevent contamination by microorganisms, even if they get beyond the epithelial surfaces, they are removed by the body's immune mechanisms, which is why the incidence of infectious diseases is low and they rarely cause illness (MURPHY, 2014).

Among the main diseases related to IAQ, Graudez and Dantas (2007) and Martinez and Callejo (2006) highlight Legionellosis and Pontiac Fever, caused by the *Legionella pneumophila* bacterium, which became known worldwide after a case in the United States.

2.9.1 Legionnaire's Syndrome

In 1976, in a hotel in the city of Philadelphia in the United States, the *Legionella* bacterium became known after an outbreak of pneumonia that affected members of the American Legion who were taking part in a convention of soldiers who had fought in the Second World War, which is why the bacterium was given the name *Legionella pneumophila* (TRABULSI; ALTERTHUM, 2008). It is estimated that 300 convention participants contracted the bacterium and 34 died as a result of complications from pneumonia. Epidemiological and microbiological studies carried out later revealed that *Legionella pneumophila* was present in the water tank of the air conditioning system and spread through the environment in the form of aerosols (MARTINEZ; CALLEJO, 2006).

Legionellosis, when it manifests itself in a milder form, is known as Pontiac fever, while with more acute symptoms it is considered a type of pneumonia. *Legionella pneumophilia* is a gram-negative bacterium that proliferates in cooling towers and can infect people who stay within 3.2 kilometers of the source of contamination. It has a lethality rate of up to 15% of identified cases. In the United States, it is estimated that *Legionella* causes 4,000 deaths a year. Cases such as these have highlighted the importance of formulating laws and standards to guarantee IAQ (GRAUDEZ; DANTAS, 2007).

2.10 Legislation and IAQ

The World Health Organization (WHO) created the first standards for indoor air quality in the 1980s (SCHIRMER. *et* al. 2011).

In 1962, the Brazilian Association of Refrigeration, Air Conditioning, Ventilation and Heating (ABRAVA) was created in Sao Paulo. Over the years it has developed a number of initiatives, and in 2008 the National Department of Indoor Air Quality (DNQAI) was created to promote IAQ through awareness-raising actions, training, forums, technical standards, scientific articles and programs (ABRAVA, 2013).

In 1973, the first ventilation standard was developed by the *American Society of Heating, Refrigerating and Air-Conditioning Engineers* (ASHRAE), called Standards for Natural and Mechanical Ventilation, giving minimum acceptable quality in outdoor environments.

With the increase in cases of EDS in Brazil, the Ministry of Health published Ordinance 3523 on August 28, 1998, with the aim of protecting the health of occupants of artificially air-conditioned environments, through preventive measures such as cleaning air-conditioning systems, establishing air quality standards according to contaminants of a physical, chemical and biological nature (GIODA; NETO, 2002).

In 1999, Ordinance 1.339 established that workers who suffer from lung disease due to air conditioning and humidification systems may be taken off work by the INSS (National Institute of Social Security), and are included in the list of work-related diseases caused by living infectious micro-organisms and parasites in group IX of the ICD-10 - J67.7 (MINISTÉRIO DA SAÙDE, 1999).

It was in 2000, on October 24, that the National Health Surveillance Agency (ANVISA) created Resolution 176, which regulated indoor air quality (SCHIRMER. *et al.* 2011).

Out of concern for the health, safety and well-being of people occupying indoor environments, ANVISA defined technical guidelines for IAQ in artificially air-conditioned environments for public and collective use through Resolution RE No. 9 of January 16, 2003 (CARTAXO. *et al.* 2007).

The aim of Resolution No. 9 was to revise ANVISA Resolution 176/00, updating it in line with the reality of knowledge in the country and making available information and experiences acquired during the first two years of RE 176's validity. It also aimed to inform the public about the importance of IAQ in artificially air-conditioned environments and its implications for human health, as well as to equip the professionals responsible for monitoring IAQ through inspections and in the planning, preparation, analysis and execution of physical projects (ANVISA, 2003).

In addition to determining conditions for IAQ, ANVISA Resolution No. 9 establishes methodologies for evaluating gases, temperature, relative humidity, aerodispersoids and bioaerosols. More broadly, since 1978 the Ministry of Labor and Employment (MTE) has established Regulatory Standard (NR) 15 with procedures related to occupational health and safety through Ordinance 3.214 for analysis in industrial environments, establishing tolerance limits for exposure to air contaminants of chemical and mineral origin. In the absence of NR-15, international standards such as the *National Institute for Occupational Safety and Health* (NIOSH) and ASHRAE can be adopted, which establish standards and maximum values for occupational exposure to air contaminants (SCHIRMER. *et al.* 2011).

With regard to microbiological contamination, Resolution RE No. 9 (ANVISA, 2003) establishes a Maximum Recommended Value (MRV) for various contaminants. For example, for fungi the MRV must be less than 750 CFU/m^3 , for a ratio of $I/E \leq 1.5$, where I represents the amount of fungi inside the environment and E shows those outside the environment. This method assesses the epidemiological tendency of the external environment and its amplification in enclosed spaces and does not accept the presence of pathogenic and toxigenic fungi inside artificially air-conditioned environments (ANVISA, 2003).

In 2003, ABRAVA created Normative Recommendation (RN) 02, with the aim of achieving indoor air quality by establishing guidelines for the design, operation and maintenance of comfort conditioning equipment. RN 02 includes technical guidelines contained in ANVISA Ordinance No. 3523/98, NBR 13971/1997 on scheduled maintenance of refrigeration systems, NBR 14679/2001 on sanitizing services, ANVISA Resolution No. 9 and CONAMA Resolution No. 3/1990 (ABRAVA, 2003).

This legislation has gained greater recognition with Law No. 13,589 (BRASIL, 2018), a federal law approved by Congress and sanctioned by the President of the Republic, making it mandatory to implement a PMOC, using the parameters of Resolution No. 9 (ANVISA, 2003) as well as the recommendations of the Brazilian Association of Technical Standards (ABNT) for all buildings for collective use and not just those with systems above 60,000 BTU/h. The law passed in January 2018 increases inspections aimed at the health of indoor occupants by raising awareness of the importance of maintaining air conditioning systems (GODINI, 2018).

Other laws and technical standards relating to indoor air quality are listed below:

- ABNT/CB-055 Brazilian committee for refrigeration, air conditioning, ventilation and heating.
- ABNT NBR 16401-1/2008 - Air conditioning installations - Central and unitary systems. Part 1: Installation projects.

- ABNT NBR 16401-2/2008 - Air conditioning installations - Central and unitary systems. Part 2: Thermal comfort parameters.
- ABNT NBR 16401-3/2008 - Air conditioning installations - Central and unitary systems. Part 3: Indoor air quality.
- ABNT NBR 15848/2010 - Air conditioning and ventilation systems - Procedures and requirements for the construction, renovation, operation and maintenance of installations affecting indoor air quality (IAQ).
- ABNT NBR 14679/2012 - Air conditioning and ventilation systems - Execution of cleaning services.
- ABNT NBR 13971/2014 Refrigeration, air conditioning, ventilation and heating systems - Scheduled maintenance.

3 WORK DEVELOPMENT

A bibliographical review, procedures and questionnaires were presented to characterize the Sick Building Syndrome in employees working in the library of the Maringà campus of the State University of Maringà.

The methods and questionnaires used to approach the workers were described.

3.1 Characterization of the Study Area

The selected study area is located in the city of Maringà, which has an area of 487.930 km^2 , located at geographic coordinates 23° 25' 30"S and 51° 56' 20'0, north of Paranà, as shown in Figure 2. It is four hundred and thirty-six kilometers from the capital Curitiba, has a subtropical climate and an estimated population of 397,437 (CENSO, 2010), making it the third largest city in the state.

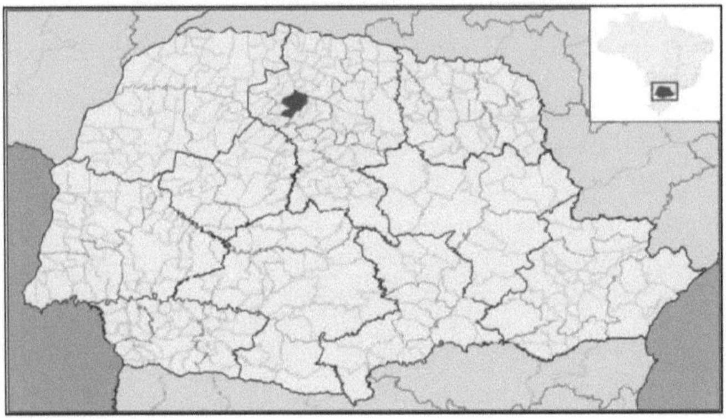

Figure 2: Location of the city of Maringà in the state of Paranà.

Source: WIDIPÉDIA, 2018.

The study took place at the Maringà campus of the State University of Maringà (UEM) between June and August 2018. The institution was founded in 1969, and expanded over the years to its current form (Figure 3), currently having around 15,000 face-to-face students and 45,000 distance learning students.

The study site chosen was the library, based on the criteria that it is a place where people circulate a lot, also because it is a space with little external air exchange in which ventilation by air conditioners is extremely important, the presence of workers and customer service was also taken into account.

Figure 3: Maringà State University Campus, Maringà-PR.

Source: GOOGLE MAPS, 2018.

3.1.1 Library

The UEM Central Library is located on the Maringà campus and was created in 1974, initially with 1,050 m^2 , having undergone various expansions over time. It currently has an area of 13,298.03 m2 covering the basement, ground floor and two more floors, the interior of which can be seen in Figures 4, 5, 6, 7 and 8. The library provides services to students and the community, with opening hours from 7.30 a.m. to 7.30 p.m..

10:30 p.m., with Mrs. Marisa Nogami as the person in charge, who contributed to this research.

Addressing this issue is interesting because of the potential risk to people's health, especially workers, who spend eight hours a day there and may suffer from allergic and respiratory problems due to the presence of dust mites and fungi that tend to inhabit the books and proliferate under the influence of the temperature and humidity, which are maintained mechanically and can be switched off at certain times of the day to preserve the books and to save energy, varying the operating time between summer and winter.

For Brickus and Neto (1999), Coutinho, Ramos and Bortoletto (2002), researchers have highlighted biological agents that are harmful to people's health, but which can also cause problems with the deterioration of works of art and literature. The humidity and temperature of the environment contribute to the development of micro-organisms that act on materials such as paper. Other agents of deterioration include insects, fungi, rodents, lighting, dust, ozone gas and environmental pollution caused by burning fuels.

Information on the power of the library's air conditioning systems was obtained from the UEM Library administration, as shown in Table 2. It was found that the equipment has power ratings ranging from 1.5 to 5 TR (ORFRINI, 2016), which is in line with Resolution No. 9 (ANVISA, 2003), establishing that from 5 TR or 60,000 BTU/h, establishments must have a qualified technical manager who will develop the Maintenance, Operation and Control Plan (PMOC) in accordance with the guidelines of Ordinance No. 3523 (MINISTÉRIO DA SAÙDE, 1998).

Table 2: Characteristics of the air conditioning system.

Location	Brand/Type	Power in BTU/h	Power in TR
1st Floor Secretarla	Philco - Split	24.000	2
1st Floor Management	Austin - Split	18.000	1,5
1st floor Meeting room	Springer - Window	30.000	2.5
1st Floor Laboratory	Hitachi - Floor ceiling	24.000	2
1st Floor Laboratory	Hitachi - Floor ceiling	24.000	2
1st Floor Commuter Room	Springer - Window	30.000	2,5
1st Floor SDI REC	Springer - Window	30.000	2,5
1st Floor Book lending	Springer - Floor ceiling	21.000	1,75
2nd Floor Study Room	Austin - Split	18.000	1,5
3rd Floor FDE	Rheem - Ceiling tiles	36.000	3
3rd Floor FDE	Philco - Split	24.000	2
3rd Floor PTE	Austin - Split	60.000	5
3rd Floor PTE	Austin - Split	60.000	5
3rd Floor Restoration	Austin - Split	18.000	1,5
3rd Floor Restoration	Springer - Floor ceiling	30.000	2,5

Source: UEM, 2018.

Figura 4: **Library administration, natural ventilation through windows and air conditioning - 3rd floor.**

Source: AUTHOR.

Figura 5: **Library administration, air conditioning unit - 3rd floor.**

Source: AUTHOR.

Figura 6: **Collection, natural ventilation by fans and windows - 2nd floor.**

Source: AUTHOR.

Figura 7: **Disposal of the book collection.**

Source: AUTHOR.

Figura 8: Periodicals, natural ventilation by windows and fans and artificial ventilation by air conditioning - Ground floor.

Source: AUTHOR.

3.2 Work methodologies

3.2.1 Questionnaire

Authorization was sought from the Library Coordinator to allow information to be obtained by visiting and questioning staff without identifying them.

The questionnaire was completed individually by the workers in order to find out the perception of the users of the environment with regard to indoor air quality. The questionnaire is an important source of information for characterizing Sick Building Syndrome in the environment.

The questionnaire was distributed to employees on July 19, 2018, based on objective and discursive questions. The interviewees indicated their profile, how long they had worked in the environment, information such as their health history and their perception of complaints related to IAQ, all of which was important for the study.

Before answering, they were given a consent form in which the employee authorizes the use of the information for the development of the research results.

The questionnaire was adapted (KAVGIC, M. *et al.* 2008) by referring to the personal characteristics of the occupant with a history of health problems, the characteristics of the location and the individual assessment of the IAQ conditions and quality according to the model in Appendix A.

4 RESULTS AND DISCUSSIONS

4.1 Questionnaire analysis

The first approach to the research was through a questionnaire on printed paper, applied to library workers on July 19, 2018, with the aim of assessing each employee's perception of QAI and analyzing the factors that may contribute to the manifestation of malaise or health problems related to low QAI.

The library has 25 employees, with 03 people working in the Administrative sector, located in rooms on the first floor. Another 12 people work on the first floor, taking care of students, organizing books and cleaning them, 03 workers on the second floor, responsible for the collection, and finally 07 people on the top floor, in the Purchasing, Administration and Restoration sectors, carrying out various activities such as cataloguing literary works, searching for new copies and general management. We can then separate 15 workers in the Operational sector, with routines such as organizing and cleaning shelves, serving students and collecting books, moving intermittently through all the library sectors and 10 in the administrative environment and cataloguing literary works (Figure 9).

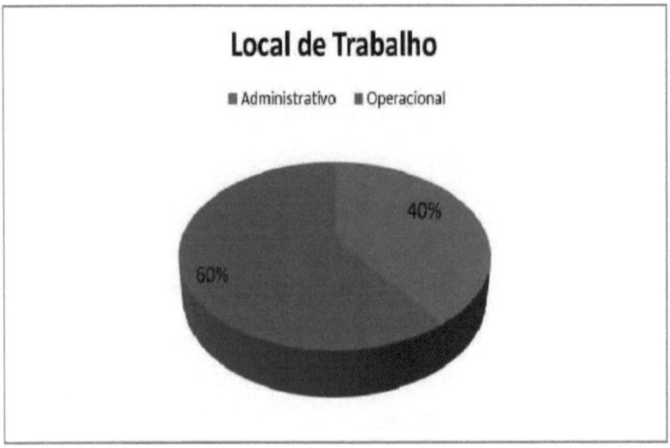

Figure 9: Distribution of employees by sector.

Source: AUTHOR.

The questionnaire had the spontaneous participation of 14 employees, corresponding to 56% of the total (Figure 10), all of whom signed a consent form and committed to the information provided, five employees are away on vacation or work issues and another seven were not on the library premises when the questionnaire was administered.

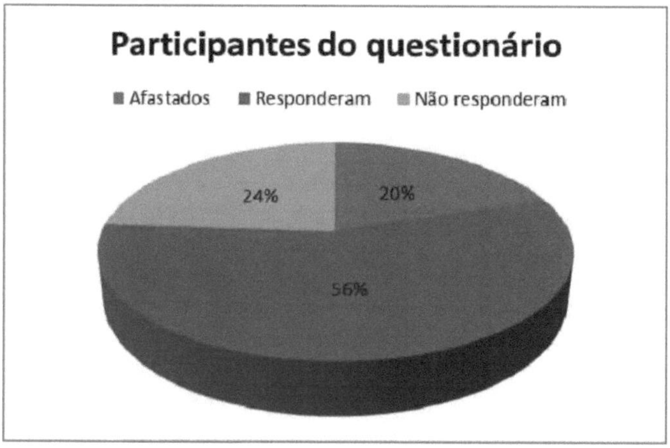

Figure 10: Employee participation in the questionnaire.

Source: AUTHOR.

Among the respondents, 57% were female and 43% male (Figure 11), aged between 27 and 64 (Figure 12) and all of them said that they worked in the library complex for 8 hours a day.

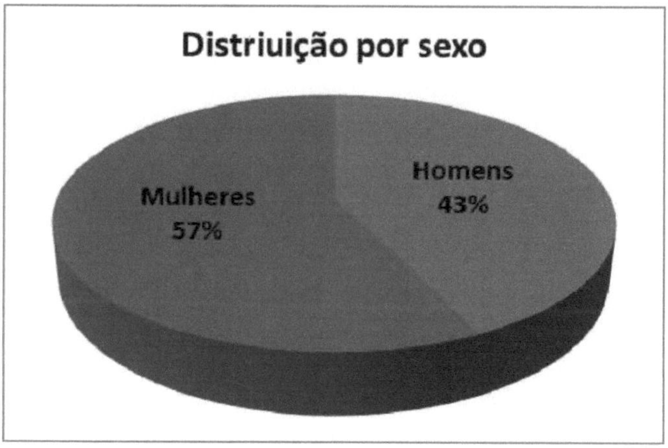

Figure 11: Employees' participation in the gender questionnaire. Source: AUTHOR

Figure 12: Employees' participation in the questionnaire in terms of age group. Source: AUTHOR.

Only one respondent said he was a smoker and when asked about health problems related to the respiratory system prior to working in the library (Figure 13), 85% said they had not had any diagnoses, while 15% said they had a history of health problems such as rhinitis, sinusitis and bronchitis, with a negative result for asthma (Figure 14).

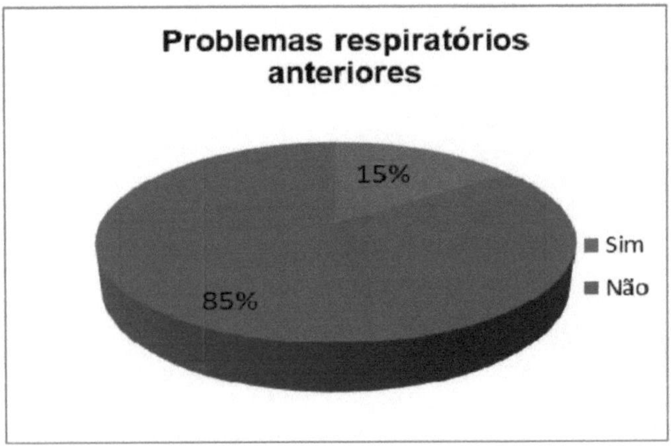

Figure 13: Percentage of people with a history of health problems related to the respiratory system before working in the library.

Source: AUTHOR.

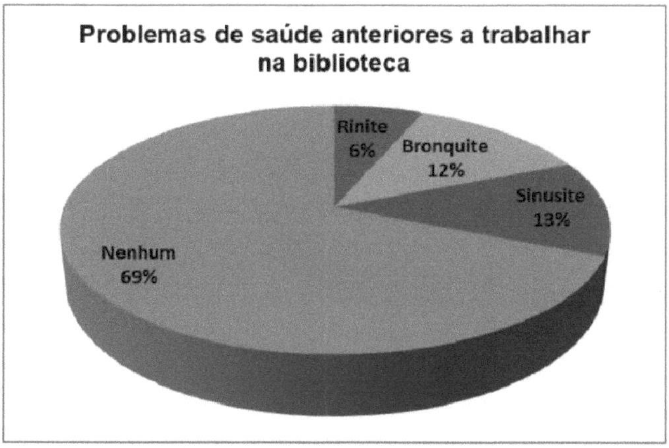

Figure 14: Health problems related to the respiratory system, before working in the library.

Source: AUTHOR.

The library can also be divided into three floors which have different characteristics: the first and third have natural ventilation through doors and windows and artificial ventilation through air conditioners; the administrative area on the first floor has artificial ventilation through air conditioners, but no windows. The second floor uses natural ventilation through doors and windows and fixed fans, and is the only room that does not have an artificial ventilation system.

The employees answered whether or not they were near sources of ventilation and it was found (Figure 15) that 43% were far from any source of ventilation, 22% near air conditioning vents, 21% near both windows and air conditioning vents and finally 14% spend most of the day near windows. This question highlights the criticism made by the interviewees about the inefficiency of the ventilation system, especially on the second floor and in the corridors where book organization activities are carried out.

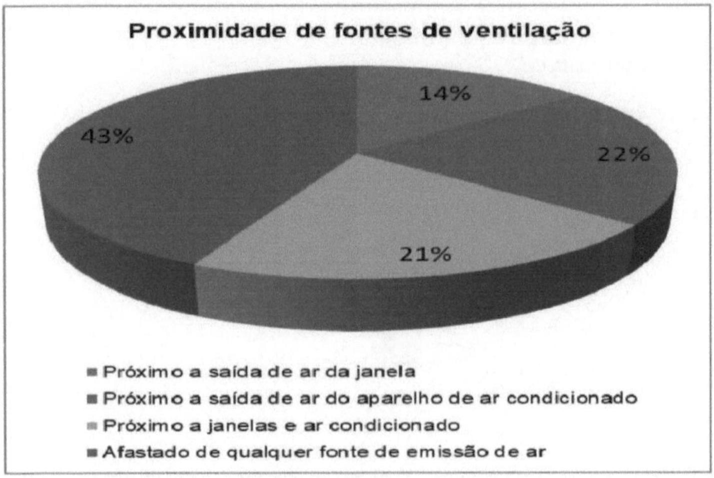

Figure 15: Permanence of respondents regarding proximity to ventilation sources.

Source: AUTHOR.

As for complaints about air quality and thermal comfort, 57% of respondents said they had already complained to the library's management or to UEM's occupational physician (Figure 16). Because there is no uniform ventilation system throughout the building, employees complain and demand that action be taken to improve it, especially on the second floor, where they have already complained about not wanting to stay.

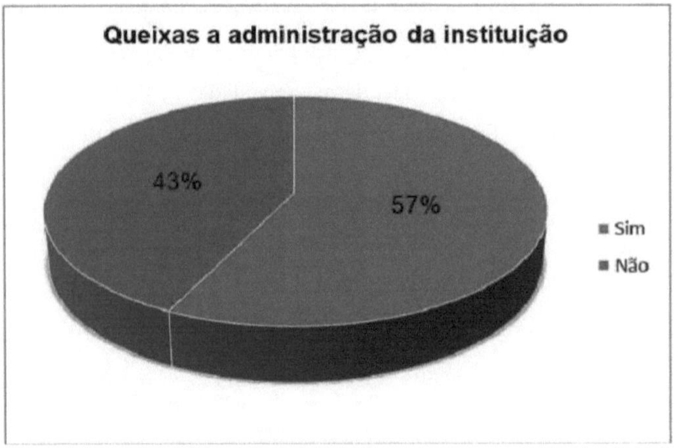

Figure 16: Workers' complaints about thermal comfort conditions and IAQ to the institution's management.

Source: AUTHOR.

When the comfort criterion is temperature and humidity, only 30% said they were satisfied with the conditions in the rooms, while 70% complained of suffering some kind of discomfort from the heat and low humidity (Figure 17), which in some way affects workers' productivity. The workers on the first and second floors complained about the cold during the winter, but occasionally the workers on the second floor complained severely about the excessive heat in the summer, as there is no air conditioning system and the fans don't meet the need.

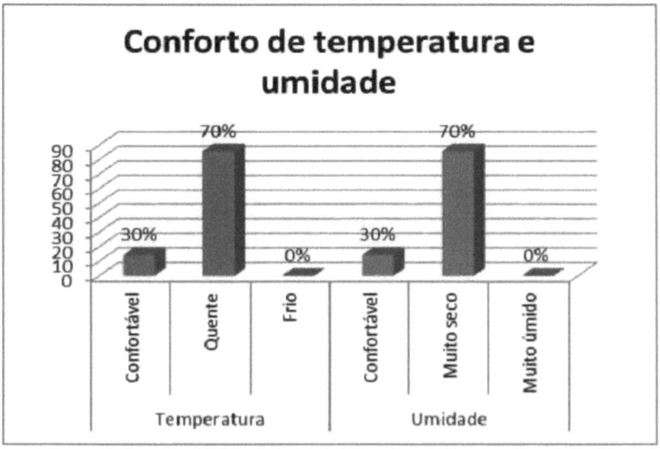

Figure 17: Workers' opinion of the library's thermal comfort. Source: AUTHOR.

Figure 18 assesses respondents' perceptions of IAQ in general, showing that 57% of workers classify indoor air as bad, 29% as good and 14% as poor, giving a 71% disapproval rate. It can be seen that the percentage of people complaining about the temperature and humidity is the same as those who disapprove of the air quality, which means that the respondents also relate the low IAQ to thermal comfort.

Figure 18: Respondents' opinion of the air quality in the library facilities.

Source: AUTHOR.

One of the main objectives of the questionnaire was to ask library workers if they had any of the symptoms related to Sick Building Syndrome (SBS), and to indicate how many they thought they felt during the working day. Based on the responses (Figure 19), 85% of people said that they experienced at least one of the symptoms of EDS. Taking into account the WHO guidelines, it can be said that the library qualifies as a sick building, exceeding the 20% determined by the WHO. Each respondent had an average of three of the nine symptoms presented in the questionnaire. The most popular were dry eyes (09 votes), itchy nose (08 votes), blocked nose and dry or irritated throat (07 votes) and sneezing (06 votes). The other symptoms received few votes, with two workers answering that they had none of the symptoms related to EDS.

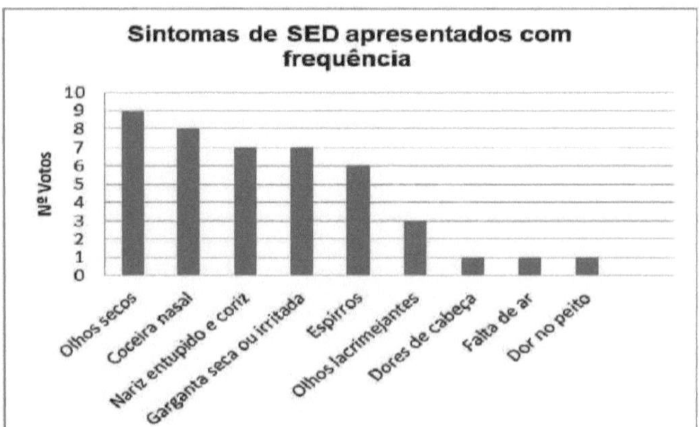

Figure 19: Number of symptoms related to EDS.

Source: AUTHOR.

When the Administrative and Operational sectors are separated, there are some inversions of symptoms between the highest votes (Figures 20 and 21), dry eyes (08 votes) is the most voted for in the Operational sector, while in the Administrative sector the biggest complaint is nasal itching (02 votes), but due to the low number of respondents to the questionnaire in the Administrative sector, it was not possible to further explore the comparisons in terms of the symptoms presented. It can be seen that in the Operational sector there is a variation in the areas affected, such as the eyes, nose and throat.

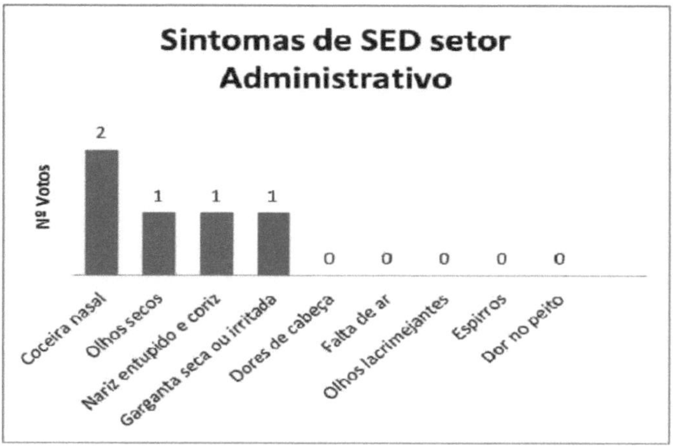

Figure 20: Number of votes for the occurrence of EDS symptoms in the Administrative sector.

Source: AUTHOR.

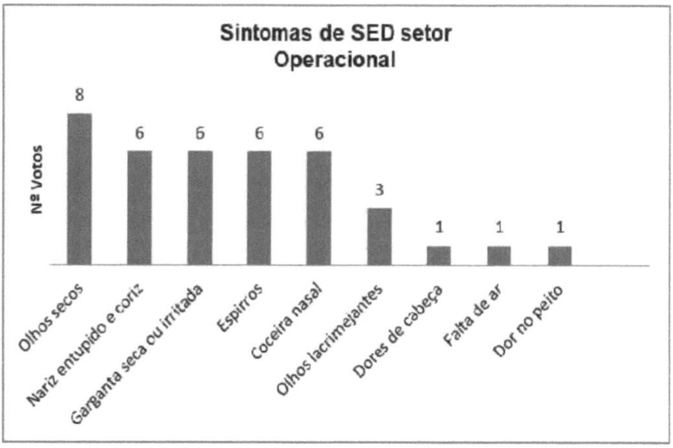

Figure 21: occurrence of EDS symptoms in the Operations sector. Source: AUTHOR.

According to Pignatari (2011), nasal obstruction can be caused by changes in temperature, the presence of allergic particles and impurities in the air, which can trigger allergic rhinitis, which obstructs the nasal concha causing itching and sneezing, which is in line with the symptoms pointed out by respondents in the Operational sector, as shown in Figure 21.

The questionnaire template can be found in Appendix A. The last phase was based on an open question for the participants to express their criticisms regarding comfort, IAQ and other conditions. The most relevant responses were:

- Improve the hygiene of the library's rooms due to dust from books, and there is a lack of manpower to meet this need;
- Improve ventilation inside the building, as there are many areas without fans or air conditioners;
- The institution should pay attention to the Bibliographic Collection sector, since the large accumulation of books leads to the proliferation of dust and fungus, damaging indoor air quality;
- Improve ventilation in the Archives sector, especially near the corridors. A worker complained that during the summer the heat is excessive, and he often gets sick and feels dizzy and faint;
- There is a high concentration of dust, especially on the second floor, which can be seen under the tables and books;
- Lack of positioning by the institution to meet the requests for improvements already demanded by the workers.

The questionnaire revealed a large number of complaints from respondents about working conditions, especially on the second floor, where there is no air conditioning system and the windows are insufficient to maintain minimum thermal comfort, which could be a determining factor in the increase in dust concentration. The UEM should therefore study the feasibility of implementing an efficient air conditioning system that not only improves thermal comfort, but also reduces the generation of dust in the environment.

The EMU suggests that a more careful approach should be taken with library staff. Given that 85% of the respondents said they had symptoms related to EDS, the institution's occupational health department should be notified to analyze the most serious cases, carrying out appropriate follow-up, helping to reduce symptoms and complaints and thus increase worker productivity.

With regard to improving the hygiene of the rooms, a problem that was very much emphasized by the workers, the accumulation of dust in the rooms is visible; Figures 22 and 23 illustrate the dirty fans.

When the workers were asked about the cleaning routines for the air conditioning and fan systems, they said that they take place randomly, without a set schedule, and are most often requested by the employees themselves when the situation becomes visibly critical. Resolution No. 9 (ANVISA, 2003) deals with the minimum frequency of cleaning and maintenance of the components mentioned in Table 1, in the context of interventions that require monthly and six-monthly intervals, covering the external air intake, filter units, condenser trays and coils.

Figure 22: Condition of the fans.

Source: AUTHOR.

Figure 23: Condition of the fans.

Source: AUTHOR

However, there were other critical points found in the environments, such as on the third floor where pigeon nests were found in the external area of the building close to the air intake of the appliances, which could contaminate the interior of the building with the fungus *Cryptococcus neofirmans* (CARTAXO. et al. 2007), as can be seen in Figure 24 and 25.

Figure 24: Pigeon nests near the equipment's air intake. Source: AUTHOR.

Figure 25: Pigeon nests near the equipment air intake on the third floor.

Source: AUTHOR.

Based on the case study of the Manguinhos Library (COUNTINHO; RAMOS; BORTOLETTO, 2002**),** UEM should create actions to continue monitoring IAQ, in order to avoid situations similar to the one that occurred in Manguinhos, so the following measures are suggested:

- Adjustments and repairs to air conditioning systems;

- Reduce the relative humidity of indoor air caused by rain by installing gutters to collect rainwater;
- Seal doors and windows, preventing air from entering that is not purified by the air conditioning equipment's filters;
- Periodically clean the literary collection mechanically, using a vacuum cleaner with a water filter;
- As for surfaces such as tables and chairs, sanitize them with an antifungal product with low toxicity;
- Maintain indoor air temperature and humidity at acceptable levels for thermal comfort and avoid large fluctuations throughout the day;
- Training for employees in good hygiene practices, as well as knowledge of how to use personal protective equipment;
- Carry out periodic assessments and implement an air quality control program.

5 CONCLUSION

With the work carried out and considering the results obtained, it was possible to question the workers and evaluate the factors surrounding IAQ in relation to biological agents, from which the following conclusions can be drawn:

- The questionnaire showed that 85% of employees had at least one symptom related to EDS, much higher than the 20% that would already characterize the building as unhealthy. It is therefore recommended that greater attention be paid to workers' health and to IAQ;
- The second floor is the place with the most specific complaints, with the absence of an efficient air conditioning system, natural ventilation is insufficient to guarantee minimum conditions of comfort for the workers;
- It was found that there is no effective maintenance plan for sanitizing the air conditioning systems, which does not respect the periodicity and recommendations of ANVISA. It is suggested that a qualified professional draw up the Maintenance, Operation and Control Plan (PMOC) in accordance with Ordinance No. 3523 (MINISTÉRIO DA SAÙDE, 1998).
- Sanitizing and maintaining ventilation systems not only contributes to people's health but is also important for preserving literary works, especially in terms of controlling temperature, humidity, dust, fungi and mold.
- It is hoped that this work will inspire further research and, as the variables that affect indoor air quality are wide-ranging, it is suggested that studies be extended, based on the legislation and standards surrounding IAQ, delving into the quantitative environmental parameters that were not covered in this study.

REFERENCES

ABRAVA. **Brazilian Committee of Refrigeration, Air Conditioning, Ventilation and Heating**. Available at: <http://abrava.com.br/?p=4123>. Accessed on: March 20, 2018.

ALBERICO, A. **Overview of air conditioning**. Ansett Technology and Engineering Workbook. Sao Paulo: 2003.

BENEVOLO, Leonardo. **History of the City**. 5.ed. Sao Paulo: Perspectiva, 2011.

BRAZIL, MINISTRY OF HEALTH. **Resolution No. 1339/1999**. Available at <http://bvsms.saude.gov.br/bvs/saudelegis/gm/1999/prt1339_18_11_1999.html>. Accessed on March 28, 2018.

BRAZIL. MINISTRY OF JUSTICE AND PUBLIC SECURITY. **Law No. 13.589/2018**. Available at: < http://www2.camara.leg.br/legin/fed/lei/2018/lei-13589-4- janeiro-2018-786057-publicacaooriginal-154702-pl.html>. Accessed on May 1, 2018.

BRAZIL, ANVISA. National Health Surveillance Agency. **Resolution No. 9**. Available at <http:// www.anvisa.gov.br>. Accessed on: March 20, 2018.

BRAZIL, ANVISA National Health Surveillance Agency. **Resolution no. 176,** available at <

http://www.anvisa.gov.br/scriptsweb/anvisalegis/VisualizaDocumento.asp?ID=136&Versao= 1>. Accessed on: 21 Mar. 2018.

BOECHAT, José L.; RIOS, José L. Pollution of Indoor Environments. **Revista Brasileira de Alergias e Imunopatologia**, Sao Paulo, v. 34, n. 3, p.83-89, mai/2011.

BRICKUS, LEILA S.R; NETO, Fracisco R. A. **Qualidade do Ar de Interior e a Quimica**. Sao Paulo, 1999. Available at <http://www.brasindoor.com.br/artigo4.php>. Accessed on March 11, 2018.

CARRIER DO BRASIL. **BIOGRAPHY, WILLIS CARRIER**. Available at: < http://www.carrierdobrasil.com.br/willis-carrier>. Accessed on: March 19, 2018.

CORBUSIER, Le. **The three human establishments**. 2.ed. Sao Paulo: Perspectiva, 1979.

CARTAXO, E. F. *et al*. **Aspects of biological contamination in air conditioner filters installed in homes in the city of Manaus**. Engenharia Ambiental e Sanitària, Manaus, v. 12, n. 2, p.202-211, dec/2007.

CORRÊA, Jorge Emanuel. **Refrigeration and Air Conditioning Workbook**. Belém: Federal University of Parà, Institute of Technology, 2010.

COSTA, M. F.; COSTA, M. A. Indoor Air Quality and Human Health. **Revista de Gestao Integrada em Saùde do Trabalho e Meio Ambiente**, Sao Paulo, v 1, n°2, article 5 - dec/2006.

COUTINHO, Eliana; RAMOS, R. M.; BORTOLETTO, M. E. Fungal Contamination of the Collection of the Manguinhos Library of the Oswaldo Cruz Foundation: Actions Developed for its Solution. **Revista Eletrônica de Biblioteconomia e Ciência da Informaçâo**, Rio de Janeiro, Oct. 2002. Available at: <http://www.redalyc.org/articulo.oa?id=14701403> . Accessed on: May 1, 2018.

ELGIN. **WINDOW AIR CONDITIONING**. Available at: <

http://www.elgin.com.br/institucional/produto.php?prod=NQ== >. Accessed on: March 20, 2018.

ENGELS, Friedrich. **Situation of the Working Class in England**. 1.ed. Sao Paulo: Bontempo, 2008.

FACULTY OF SCIENCE AND TECHNOLOGY, UNIVERSITY OF Coimbra. **Technological Specialization in Environmental Quality**. Manual for Environmental Microbiological Monitoring. Coimbra, 2012, 60p.

FROTA, A.B. SCHIFFER, S.R. **Manual do Conforto Térmico**. 8.ed. Sao Paulo: Ed. Studio Nobel, 2016.

GIODA, Adriana; AQUINO NETO, Fracisco R de. **Chemical Pollution Related to Indoor Air in Brazil**. Quimica Nova, Sao Paulo, v. 26, n. 3, p.359-365, 2003.

GODINI, Charles - **Law 13.589: Benefits and legal responsibilities in the application of the PMOC.** Published in March 2018. Available at: < http://www.engenhariaearquitetura.com.br/2018/03/lei-13-589beneficios-e-responsabilidades-legal-na-aplicacao-do-pmoc>. Accessed on: May 1, 2018.

GOOGLE MAPS. **Maringà, Paranà**. Available at: <

https://www.google.com.br/maps/place/Maring%C3%A1,+PR/@-24.386028,-51.725257,8.02z/data=!4m2!3m1!1s0x94ecd6ee42618c1f:0xc356592828e03925>. Accessed on: April 29, 2018.

GRAUDENZ, Gustavo Silveira; DANTAS, Eduardo; Indoor Pollution and Building-Related Diseases and Symptoms. **Revista Brasileira de Medicina**, Sao Paulo, v. 2, p. 23-31, feb/2007.

HEDGE, A; STERLING, E.M; STERLING, T.D. **Building illness indices based on questionnaire responses.** Atlanta, 1986. Available at <http://www.sterlingiaq.com/photos/1064444052.pdf>. Accessed on April 19, 2018.

KAVGIC, M. *et al*. Analysis of Thermal Comfort and Indoor Air Quality in a Mechanically ventilated theater. **Energy and Buidings,** California, v. 40, n. 1, p.1334-1343, dec/ 2008.

LAMBERTS, Roberto; DUTRA, Luciano; PEREIRA, Fernando. **Energy Efficiency in Architecture**. 3. ed. Brasilia: Procel, 2013.

MARTiNEZ, F. J. R; CALLEJO, R. C. **Edificios Saludables para Trabajadores Sanos: Calidad de Ambientes Interiores**. Valladolid: Junta de Castilla y León, 2006.

MONTAGNER, C. C.; PASCHOALINO, M. P.; JARDIM, W. F. Aplicaçao da Fotocatàlise Heterogênea na Desinfecçao de Agua e Ar. vol.4. Campinas: Universidade

State of Campinas - UNICAMP. Institute of Chemistry. **Environmental Chemistry Laboratory**, Feb/2005.

MURPHY, Kenneth. **Immunobiology**. 7.ed. Sant Louis: Artmed, 2010.

PENA, S. M. **Air Conditioning and Refrigeration Systems**. 1. ed. Brasilia: Procel, 2002.

PEREIRA, S. S. *et al.* Fungal Microbiota Control in a Private University Library and its Impact on Occupational Health: ENCONTRO LATINO AMERICANO DE INICIAÇÂO CIENTIFICA, 14., 2011, Sao José dos Campos. **Electronic proceedings...** Sao José dos Campos, UNIVAP, 2011. Available at: < http://www.inicepg.univap.br/cd/INIC_2010/anais/arquivos/0383_0320_01.pdf>. Accessed on: April 10, 2018.

PIGNATARI, S.S.N. **Nasal Obstruction**. Available at: < https://drauziovarella.com.br/entrevistas-2/obstrucao-nasal/>. Accessed on: April 10, 2018.

PINTO, A.K.; XAVIER, J.N. **Manutenção: Funçâo Estratégica.** 3.ed. Rio de Janeiro: Qualitymark, 2009.

PIRANI, Marcelo José. **Refrigeration and Air Conditioning**, Part II. Itajuba, 2001, 60p.

REOLON, Aline; PEREZ, L.R.R.; MEZZARI, Adeline. Prevalence of Cryptococcus neoformans in urban pigeons in the city of Porto. **Revista Bràs. Patol. Méd. Lab**, Porto Alegre, v. 40, n. 5, p.293-298, Oct/2004.

RIENZO, Cristiane Di. **A Memoir of Refrigeration and Air Conditioning in the Workplace**. 1. ed. Sao Paulo: Leograf, 2006.

RUAS, Alvaro César. **Thermal Comfort Assessment: Contribution to the Practical Application of International Standards**. 1. ed. Sao Paulo: Spel Gràfica e Editora, 2001.

SCHAECHTER, Moselio. *et al.* **Biology: Mechanisms of Infectious Diseases**. 3.ed. Rio de Janeiro: Guanabara Koonga S.A, 2002.

SCHIRMER, Waldir N. *et al.* Indoor Air Pollution and the Sick Building Syndrome. **Revista Ciência e Saùde Coletiva**, Rio de Janeiro, v. 16, n. 8, p.35833590, nov/2011.

SILVA, Péricles. **Architectural Acoustics & Air Conditioning**. 4.ed. Belo Horizonte: EDTAL E.T Ltda, 2002.

SPENGLER, J. D.; SAMET, J.M.; MC CARTHY, J.F . ***Indoor Air Quality Handbook***. 1.ed. Washington: McGraw-Hill, 2001.

SPRINGER CARRIER. **Products.** Available at: <http://www.springer.com.br/pt/produtos>. Accessed on: March 18, 2018.

STERLING, Theodor D; COLLETT, Chris; RUMEL, Davi. The Epidemiology of Sick Buildings. **Revista Saùde Pùblica**, Sao Paulo, v. 25, n. 1, p.56-63, 1991.

STOECKER, W.F; JABARDO, J.M.S. **Refrigeraçãoçâo Industrial**. 2.ed. Sao Paulo: Ed. Edgar Blucher Ltda, 2002.

STRUSZ, M.C.; MACHADO, J.M.; BRICKUS, L.S. Analysis of a Fungal Contamination Accident in a Public Library in the Municipality of Rio de Janeiro. **Revista Brasileira de Saùde Ocupacional**, Sao Paulo, v. 32, n. 115, p. 69-78, jan./ 2007.

TORTORA, Gerar. J; FUNKE, Berdell R.; CASE, Christine L. **Microbiology**. 6.ed. Sao Paulo: Artmed, 2003.

TRABULSI, L.R; ALTERTHUM, Flavio. **Microbiology**. 5.ed. Sao Paulo: Atheneu, 2008.

XENOS, Harilaus G. **Gerenciarando a Manutençâo Produtiva**. 1. ed. Belo Horizonte: Desenvolvimento Gerencial, 1998.

APPENDIX

APPENDIX A: Indoor Air Quality Questionnaire

Artificially

Date:	**Time:**
Name:	
1 - Respondent Gender ( ) Male ( ) Female	Age:
2 - Length of time working in the Library: ( ) less than 1 year ( ) 1 to 2 years ( ) 2 to 3 years ( ) more than 3 years	

3 - Time spent in the Library per day:

( ) less than 30 minutes ( ) from 30 minutes to 4 hours ( ) from 4 to 6 hours

( ) 6 to 8 hours

---

4 - Place of work:

( ) 1st Floor ( ) 2nd Floor ( ) 3rd Floor

---

5 - It remains close to the point of discomfort for most of the day:

( ) Next to the window vent.

( ) Near the air outlet of the air conditioning unit

( ) Away from any source of air emissions

---

6 - Before working in this environment, you have already presented:

( ) Asthma ( ) Rhinitis ( ) Bronchitis ( ) Sinusitis

( ) None

---

7 - You smoke:

( ) Yes ( ) No

---

8 - He has already complained about the air quality to the institution's administration:

( ) No ( ) Yes

9 - How would you describe the comfort of the library?

Temperature:

( ) Comfortable ( ) Warm ( ) Cold

Humidity:

( ) Comfortable ( ) Very dry ( ) Very humid

Air quality:

( ) Excellent ( ) Good ( ) Poor ( ) Bad

The noise caused by the air conditioning system:

( ) Quiet ( ) Not very noisy ( ) Very noisy

10 - You experience any of these symptoms frequently:

( ) Dry eyes	( ) Watery eyes	( ) Sneezing
( ) Headaches	( ) Stuffy nose and runny nose	( ) Itchy nose
( ) Shortness of breath	( ) Dry or irritated throat	( ) Chest pain

11 - Describe any other observations you think are relevant to the air quality in the environment where you work:

More Books!

yes

I want morebooks!

Buy your books fast and straightforward online - at one of world's fastest growing online book stores! Environmentally sound due to Print-on-Demand technologies.

Buy your books online at

www.morebooks.shop

Kaufen Sie Ihre Bücher schnell und unkompliziert online – auf einer der am schnellsten wachsenden Buchhandelsplattformen weltweit! Dank Print-On-Demand umwelt- und ressourcenschonend produzi ert.

Bücher schneller online kaufen

www.morebooks.shop

info@omniscriptum.com
www.omniscriptum.com

OMNIScriptum

MIX
Papier aus verantwortungsvollen Quellen
Paper from responsible sources
FSC® C105338

Printed by Books on Demand GmbH, Norderstedt / Germany